Growing-Up Exercises

Other Books by Marjorie Craig

Miss Craig's 21-Day Shape-Up Program
Miss Craig's Face-Saving Exercises

Miss Craig's Growing-Up Exercises

to help children grow up to be healthy adults

Random House New York

Copyright © by Marjorie Craig

All rights reserved under International and Pan-American Copyright Conventions. Published in the United States by Random House, Inc., New York, and simultaneously in Canada by Random House of Canada Limited, Toronto.

Library of Congress Catalog Card Number: 72-11418
ISBN: 0-394-48491-6

Manufactured in the United States of America
First Edition

Design: Charles Schmalz

Contents

Movement Is Life   7
The Infant   10
The Toddler   14
When the Child Is Five   20

Your Child's Body   23

The Daily Program   29

Exercises for Special Problems   73

Movement Is Life

Every day of my life, in my exercise studio and outside it, I meet adults with "bad back" and low back strains, with knock-knees, bowlegs, flat feet and swaybacks—with all the problems of posture and movement that make at least half the people in this country feel "not very well" at least half the time.

Many of these people are resigned to discomfort because their parents had the same problems and they believe they have inherited hopeless difficulties. In most situations that is simply not the case. Most adult problems of this nature have their beginnings in infancy and childhood and they <u>can</u> be avoided. Just as you teach your children to brush their teeth to insure their dental health, so you should teach them the daily habits that will insure sound bodies. Your child won't "outgrow" backaches or posture problems on his own—as you did not. Consult an orthopedic physician the moment you notice any sign of orthopedic problems in your child, whatever his age.

Then from the age of five onward, all children should do daily exercises, no matter how active they are, no matter whether you live in the country or the city, or whether the youngsters are athletic or not. When you look at your five- or six-year-old his back may seem to be very straight to you. However, bear in mind that children who don't exercise will be more likely to begin developing postural problems by the time they are teen-agers.

Your child should certainly be encouraged, from the time he is very little, to run and play, for both his physical and emotional well-being. However, it is a mistake to think that games and athletics are "enough" exercise. From the time a human being assumes the upright position, the force of gravity is constantly pulling down on his body. If you think about sports, you will realize that most games are "one-sided" or are dependent on the forward thrust of the body and limbs. For this reason, athletics—though they teach the child all sorts of wonderful things, including physical things—do not take the place of a daily program of corrective and preventive exercise.

Movement takes place through the bones and muscles, and the way a child moves determines how he will grow and develop. Though it would be foolish to prescribe a course of physical exercise for very young children, still, as we will see in the following pages, it is important to point out that even little babies are sometimes handled in ways that result in postural and growth problems.

I do not think that children less than five years old, under most circumstances, should be given formal exercises. (On page 73 there are a few exercises for very young children who have specific walking or postural problems.) However, from age five on, if you introduce the exercises in this book properly, children will enjoy and look forward to doing them every day.

The only rule is not to force the child: ease him into the program gently and with good spirits. I think you will find, as I have in my work with young children at the Columbia Presbyterian Medical Center in New York, that your children will like discovering what they can do with their bodies. At the same time, you will have bestowed upon your child the lifetime gift of a health-giving daily habit.

The Infant

Certainly, small infants shouldn't be made to follow a program of exercises. And certainly, no mother is going to hover over her baby's crib watching his every movement. But there are things you can do to prevent even the smallest baby from developing problems which will affect back, legs and posture in his adult life.

The Crib

Make sure the baby's mattress is firm and the crib is large enough to allow the infant to move around freely.

Though your heirloom bassinet may be pretty, it really isn't a good idea to confine the baby in so small a space for any length of time. When his muscles tell him to move, he should have the room to do so.

Try to vary the position of the baby in the crib. Don't turn his head in the same direction every time he's lying on his stomach. Turn him on his back some of the time so he can move his arms and legs freely.

If, when he's on his stomach, his knees are always pulled up under him and his feet turned inward, this could cause a pigeon-toed condition and knock-knees. When he's on his stomach, the knees should be bent and turned outward <u>slightly</u>. Don't pull at him too much or exaggerate the position, but occasionally and very gently adjust his legs.

The physical development of the child proceeds from the axis, or center line, of the body to the extremities. The very young child is capable of movements involving the large muscles of the arms nearest the trunk before he can make the separate movements of the wrist, hands and, finally, the fingers.

Attach a mobile or a line of rattles to the crib where he can easily see it. This will encourage him to wave his arms and reach for the toy. Soon you'll notice his legs and feet moving in excitement too.

Carrying the Baby

Don't wrap your child so tightly in a blanket that he can't move around.

Don't always carry the baby slung over your hip. This will tend to make him bowlegged, as whole villages of people are bowlegged where this is the only way babies are held.

There's no need to worry every time you lift up the baby about whether you are holding him properly. Simply vary the position from time to time; don't use any carrier or seat too much; and let the baby have as much freedom of movement as is safely possible.

In recent years many young mothers and fathers have been carrying their children in commercial slings that go over the hip and in various other carriers that are designed for use on the back or in other positions.

I suppose there is no harm in using this equipment once in a while when it is convenient. However, I don't think any type of carrier should be used too often or for very long periods of time. All these devices severely limit the movement of the child's arms and legs. Movement is life, and the less we do to restrict the small child, the better.

The other point to be made is that the hip-sling, for example, will make the parent carry one hip and one shoulder higher than the other—a very unhealthy posture for the person doing the carrying. Also, people wearing a back pack tend to round their shoulders, sink their chests and protrude their abdomens—another posture syndrome to be avoided.

Growth

Not all parts of the body grow at the same time or the same rate. They each follow a growth sequence of their own. Muscles, bones, limbs, trunk and the various organs have periods when they grow rapidly and periods when they slow down. As one part grows, another may remain stable.

Muscles should develop as the bones lengthen. If this doesn't happen the muscles may become flabby, and movements may become awkward.

Muscle tissues develop in strength and endurance through exercise. As the child grows, all of the muscle groups should be exercised equally. If they are not developed equally, the body could be thrown out of balance and strain might be placed on some parts. For this reason, you should turn small infants frequently and try to avoid games and equipment for the older child which favor the use of only one side of the body.

The shape and position of certain bones of the human body are affected by the pull of the muscles attached to them and by the force of the body weight. Thus, the movement of the body affects the way the baby's bone structure and his muscles will develop.

Physical Development

Never force your child to sit, stand or walk. When his muscles are ready, he will be ready.

If he is forced to sit before the muscles are strong enough, the muscles may be weakened and the upper spine may be thrown out of alignment. He could develop round shoulders, a sunken chest and a protruding abdomen, all of which could cause weakness throughout his life.

Don't prop pillows behind the baby's back to try to get him to sit. A slanted tray holds the baby in a better position for feeding than pillows do, but don't leave the baby in the tray too long.

There's nothing wrong with playing seesaw or other games that pull the baby up by his arms. However, don't play these games when the baby is too young; don't play them too often; certainly, don't force him further up than his muscles want to go. Watch carefully for signs of fatigue and stop before the baby is cross and obviously tired. Often small infants will be so delighted by the singing of a nursery rhyme or the fun of a game that they'll go on and on, past their own limits.

While we're discussing infant exercises let me repeat something: don't, for Heaven's sake, subject small children to a regimen of calisthenics. Babies and small children are busy enough with the exercise of learning to sit and crawl and walk—nature is running its own gym class.

Sitting

When his back can support him, the baby will sit by himself. When he's sitting on the floor, his knees should turn outward. If the thighs are always turned inward, the position could cause a disturbance in the development of the bones and ligaments of the hip socket and develop a weakness there. This joint is very important for weight-bearing. If his knees are continually turned inward, gently turn them the other way. Do not tug at his legs but make an occasional adjustment.

Crawling

First the baby will learn to turn over by himself, and you will notice that he's able to take long trips by rolling himself where he wants to go (the reason you should <u>never</u> leave a small baby unattended on an unguarded bed).

His developing arm muscles help him turn over and prop himself up. As the legs and pelvic muscles develop, the baby will be ready to crawl.

Babies learn to crawl at different ages with all sorts of exotic techniques. Don't worry if, at first, your baby isn't crawling like your neighbor's. However, if after several weeks you notice that he is favoring one side, or dragging a leg or not using one arm, the condition should be checked by a good physician. Though such crawling may signify nothing, there is the possibility that the baby's body is signaling some difficulty that should be examined.

Let your child crawl on all fours as long as he wants to, for it will help to develop his pelvic and trunk muscles. If he is forced to stand too soon and the weight-bearing muscles, ligaments and joints are not ready, they will be weakened and may not develop properly. The alignment of the legs could be affected and too much strain placed on the feet, which may not be ready to support the body weight. The body will not be balanced properly and the curves of the spine will increase if the baby is made to stand before his body is ready. "As the twig is bent, so it will grow" applies exactly to the baby's growing body.

Don't worry if your baby isn't beating the world record for age of standing. Don't be pressured by your mother or your neighbor to "help him a little." In time he will stand, and walk, and run—as we all did. When he is walking all over the house, you'll think back nostalgically to the days he was less mobile.

The Toddler

Shoes

The idea that it's "natural" for people to go without shoes doesn't hold for modern civilization. You might bear in mind that for centuries the only unshod horses have been those that graze on soft grassy plains. It's one thing to go barefoot on sand or grass. Walking without good support on pavement or hard floors isn't "natural." It tends to flatten the arches of the feet (and to pick up glass and splinters, among other things).

Children should be put in sturdy comfortable shoes soon after they begin to walk. There are all kinds of shoes on the market for small children. Avoid those that attempt to imitate grown-up styles. For example, though you may think sneakers will look cute on your twelve-month-old, who has just begun to stand and move around, they really aren't adequate footgear for him. For your little girl, don't buy Mary Janes or sandals or other party shoes until she's quite a bit older, and even then, limit their use to special occasions.

Young children need sturdy shoes that give them good support. In fact, for the very young I prefer shoes that come up over the ankle. The shoe should fit snugly around the heel and there should be plenty of space for the toes to move around.

The Developing Foot Muscles

Some small children, when they first begin to walk, tend to put their weight on the inside arches of the foot, with the ankle turned in and most of the weight on the big toe. If you notice this condition persisting, see a good orthopedist, who will probably prescribe a shoe with a wedge on the inner side of the shoe to throw the weight on the outside of the foot.

However, don't rely only on the shoes to correct this condition. The muscles themselves have to be strengthened through exercise. The following exercise will help. Approach it with your child as if it were a game, working with him only when he's feeling fresh and only for short periods at a time. Do the exercise yourself right along with him so he gets the idea and wants to imitate what you are doing.

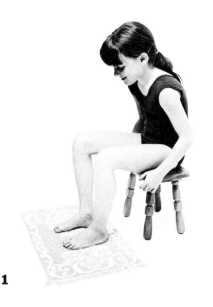  

1 **2** **3**

1. Put the child on a stool with his feet on a long Turkish towel that is spread out on the floor in front of him. His feet should be flat on the floor, four inches apart and parallel, pointing straight ahead.

2. Grasp the towel with the toes.

3. Point the toes inward, pigeon-toed, and pull the towel in toward the center.

4. Release the towel. Pull toes upward, keeping the heels on the floor.

5. Turn the toes outward to the original parallel position. Don't turn way out.

6. Grasp towel with toes again and bring more of the towel in toward center in pigeon-toed position. Release towel and repeat movements until towel is bunched up between the feet. Spread the towel out and repeat until the child tires of the game.

Building a Mountain
(For turned-in ankles)

4 **5** **6**

The child has now begun to walk at his own time, at his own pace, and you have got him in the proper shoes.

When he first begins to walk he may turn his feet outward. If this goes on too long, it may make him flatfooted by throwing the weight of the body inward onto the arches, which may then flatten and not develop properly. As the child grows, the body weight should gradually fall on the heels, the outside borders and the balls of the feet so that when he walks, his feet bend in a rocking motion. If he walks properly the muscles in the feet will pull upward and outward to help form the arches. His feet should point straight ahead when he is standing and walking.

If you notice that the new walker is constantly toeing in, or turning his feet out or moving in a flatfooted way, then exercise is called for.

As I've said before, I don't advise difficult or rigorous exercise for children under five. However, if your toddler has walking problems, here are two simple things to do. Make a game out of these exercises, telling the child what to do as if it were fun and not a chore. (Both of these are fun.)

Walking a Line
(For flatfeet)

Find a straight line on the floor. You can use the edge of a rug or a line on the floor or in the pattern of your linoleum. The line should run at least 10 or 12 feet.

Do this exercise yourself, playing follow the leader with the toddler behind you. Then turn around and have him precede you down the line, as the leader, so you can check that he's walking properly.

1. Walk with the line between the two feet and with the feet exactly parallel to the line, pointing straight ahead.

2. Walk so that the heel touches the floor first. Turn foot inward (the foot is turned so that it is parallel with the line). The outside border of the foot touches the floor next.

3. The ball of the foot and toes touch the floor last. Make sure the knees bend over the foot, not inward or outward.

Don't get so carried away with the game that you forget to watch the baby's feet. (At the same time you're helping the child's feet, you'll be helping your own.) Do a few times every day.

Picking Up Marbles
(For flatfeet)

1. Put the child on the floor; have him sit up, with his knees turned outward. Get ten marbles (and watch carefully that he doesn't put them in his mouth). Put the marbles in front of him between his legs.

2. Show him how to grip each marble with his toes.

3. Then have him pick up the marbles, one at a time, with his right toes bringing each one up to his left hand and then handing it to you. Watch that he turns his right knee outward as he brings the marbles to his left hand.

4. Replace the marbles on the floor and have him use the toes of his left foot to carry the marbles to his right hand, turning his left knee outward. Continue until the child loses interest or seems tired.

Physical Activities for Children under Five

Encourage your young child to be as physically active as possible up to the limit of his own energy. (Of course, he shouldn't play so long or so hard that he becomes fatigued.) Games using the body are fun and will help muscular development. More than that, learning to take pleasure in physical activity is important for all of a person's life. Mental alertness and physical fitness are closely connected, and the early years are the time for the child to learn the pleasure and profit of having a good body.

Swimming is one of the best sports, in childhood as well as later life. Since you don't have to work against gravity in the water, you don't overdevelop muscles, and the weaker muscles have a better chance to function than they do on land.

There are excellent classes around the country that teach swimming to very young children, with the help of the mothers. Of course, you must watch like a hawk when a small child is in the water.

The one physical activity that children of all ages seem to love and that I hate is standing on your head. The neck was not made to bear the weight of the body. It is not a weight-bearing joint like the hip. The neck encloses a vital nerve center from which the nerves branch out to enter the head and out into the arms. Injury to that area could obviously be terribly serious.

Also, standing on your head involves a very precarious balance, and a child falling from that position could easily hurt his back. Don't let your children stand on their heads—especially not young children.

When the Child Is Five

Posture

Your child's posture is one of the most important elements of his development, both physical and psychological. Posture means position, the way the body is held. Good posture will help keep the internal organs in their proper place so that they can develop and function well. Good posture will keep undue strain off the major weight-bearing joints: the lumbar spine, the hips, knees, ankles and arches.

Bad back, sciatic nerve pain, sacroiliac and lumbosacral strains are common and painful adult complaints which may be avoided by helping your child attain good posture early in life. If he acquires it when he is very young, good posture will stay with him for the rest of his life.

The way the child sits and stands not only reflects how he is feeling but it also influences how he feels and how other people react to him. The child who stands up straight looks alert. He is more likely to be sought after and feel sure of himself than the hunched-over child would be. A child who is shy and bashful and whose posture reflects his lack of self-confidence will not be "cured" by exercise alone. You should give him a feeling of understanding and of being well loved; this will give him confidence in himself, which will show in the way he stands and moves.

To achieve and maintain good posture, the muscles of the body must be kept in good tone so that it takes little effort to carry the body properly and the muscles do not need to strain and tense unduly.

Muscle tone is achieved through exercise. Your child should be encouraged to run and play as much as possible, but that is really not enough for excellent muscle tone. Gravity is constantly pulling down on everything, including the human body, and, as I have mentioned, most games involve forward motion rather than exercise that counteracts the pull of gravity.

Overweight and Posture

Overweight is an important cause of bad posture. Excess weight puts undue strain on both the muscles and the weight-bearing joints of the growing child. An overweight child is more easily fatigued by physical exercise and will tend to slump when he is tired. This physical condition will have a psychological effect, and the result will be an unhappy, unhealthy child.

Deep Breathing

Deep breathing is essential to a properly developing and functioning body. It is important to expel all the carbon dioxide from the body and to fill the lungs with fresh oxygen to pass into the bloodstream.

Children are often taught deep breathing, but almost as often they are taught incorrectly. Breathing should not originate from the stomach but should come from the breathing muscles in the chest: the intercostal muscles and the diaphragm.

You can begin deep-breathing exercises with very young children and can instill this health-giving habit very early in the child's life.

The movements of breathing are easy to understand: when you breathe in, the rib cage lifts and expands, the diaphragm goes down, the chest cavity is enlarged, and the lungs expand as they fill with air. When you breathe out, the rib cage relaxes and the diaphragm comes up, forcing the air out of the lungs.

Get your five-year-old to practice deep breathing for a few times every day and increase the time as he grows older until he is taking about twenty deep breaths when he is ten to twelve years old.

When your child is still young, tell him this exercise is like filling a balloon inside him:

1. Place hands on rib cage.

2. Slowly blow out all the air in the lungs, to empty the balloon.

3. Breathe in a long, slow breath.

4. Now blow out, slowly and as long as possible.

Do 5 times, for a young child; about 10 times for an eight-year-old; and 20 times from ten years old on.

Exercises Not to Let Your Children Do

This is a slightly tricky subject because you may find that your child's school is giving him exercises which aren't really good for him. If that is the case, have a talk with his teacher or with the school administration, or failing that, try to get other parents to join you in your concern.

The same advice holds for strenuous sports which, in some communities, are set up for very young children. I would not recommend that boys under fifteen play tackle football, since serious injuries may take place.

I also believe that children should not be given "Atlas-type" exercises that build bulging muscles and stress weight-lifting and heavy pulling.

Recent studies have shown that exaggerated knee bends, such as the full squat and "duck walk," can damage the muscles and cartilage that support the knee. This could cause knee injuries later in life.

Do not do any exercises which involve straight leg lifts with both legs at one time. The pelvis is tilted forward in this exercise and strain is placed on the lower back. When you are lifting one leg straight up, always bend the other leg at the knee.

Only certain kinds of sit-ups are good for you. Do not raise the body from the lying position with the hands behind the neck and the legs stiff. Here, again, the pelvis will be tilted forward and strain placed on the lower back. If you do sit-ups, see the method on page 40, where the knees are bent, or consult my 21-Day Shape-Up book* for adults, where (Exercise 24, p. 60) the arms are extended back over the head. Whichever method you use, make sure the knees are bent so the pelvis is not tilted forward.

Don't ever do backward bends. They stretch the stomach muscles and overdevelop the back muscles.

Toe touches with stiff knees should be avoided. Do not bend forward from the waist with stiff knees. This places strain on the lower back muscles and ligaments.

Don't do side stretches with the knees stiff and the pelvis tilted forward. These are very common in school programs and should be done only if knees are bent and hips tucked under.

As we have already warned, don't ever stand on the head. The neck was not made to support the weight of the body and real damage can be done. If, like a yogi, you want to get more blood to your head or your child's, invest in a slant board or use an ironing board carefully propped in a slanted position.

*21-Day Shape-Up Program for Men and Women (New York: Random House, 1968).

Your Child's Body

Before you start your child on a daily program of exercise you should know something about the structure of the body.

The Spine

The adult spine has four curves, the cervical, thoracic, lumbar and sacral; a newborn infant has only two curves, the thoracic and the sacral. The cervical curve, at the neck, starts to form as the infant begins to hold up his head. The lumbar curve appears when he begins to stand.

When the child first stands, the pelvis tilts forward and the stomach protrudes. If he develops properly, his weight shifts so that the body straightens and the pelvic girdle tilts backward, pulling the stomach muscles into correct position.

The spine rests on the pelvic girdle. It supports the head and the rib cage. If the spine is not held in good alignment, the internal organs as well as the bone structure are affected; weight-bearing joints will be strained; and antigravity muscles will become weakened from insufficient muscular activity. When the spine is held properly, all of the other segments—the head, rib cage and pelvis—will fall into place. These three segments must be properly balanced at all times. If they are not, as we will see, postural faults will occur.

The pelvic girdle must be balanced and held correctly, for not only will it affect the alignment of the spine above it, causing lordosis (sway-back), but it will also affect the position of the legs and result in knock-knees, bowlegs or flatfeet.

The rib cage is attached to the spine, and the shoulder girdle rests on the rib cage. The shoulder girdle is made up of the

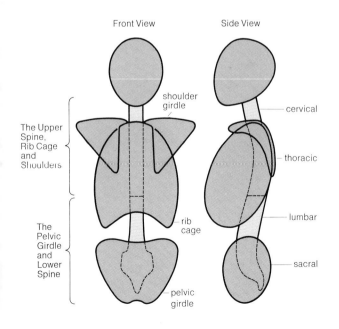

collarbone (clavicle) in front, and the shoulder blades (scapulas) in back. The shoulder itself is made up of the prominence of the upper arm, the scapular region (region around the shoulder blades), the upper part of the chest and the armpits.

The rib cage must be held correctly, for it affects the position of the head and shoulders. Incorrect position of this segment will cause kyphosis, where the head and shoulders sag forward, the chest becomes sunken, the back rounds, the muscles of the chest and back become weak, and the shoulder blades stick out like wings. If this condition persists as the child grows, the movement of the arms may become limited because the shoulders are pulled forward.

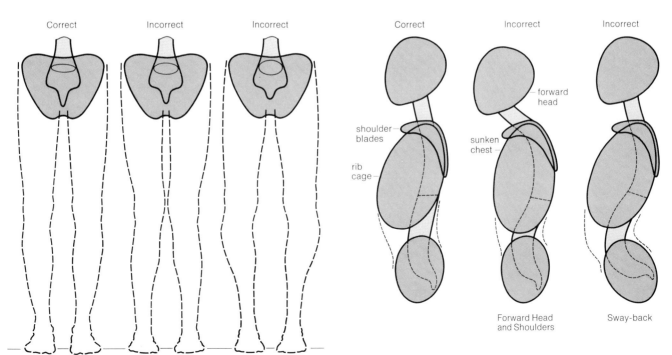

Antigravity Muscles

These are the deep muscles of the back which extend from the base of the spine up to the head. They hold the body in sitting and standing positions.

Gravity is constantly pulling downward and these antigravity muscles are always in use to hold the body upright against that downward pull. They act on the spinal column, head, ribs and pelvis. They hold the spine upright, bend the spine from side to side, rotate the spine as in twisting, and move the head.

These antigravity muscles are important in connection with many disabilities that arise from faulty posture, back injuries, and muscle strain. If these muscles aren't sufficiently exercised, the child will have to work too hard to hold his body upright.

The Abdomen

All muscles have tone or firmness. Without good tone, muscles sag and lose their usefulness and their beauty. Tone has an especially important influence when the muscles are in the resting position.

The stomach muscles control the position of the abdominal viscera with their normal tone. If the stomach muscles are in good tone, they will, even in their resting position, hold the internal organs in their proper place. If they are not in good tone, they will sag and the abdominal viscera will also tend to sag forward, possibly causing internal as well as external disorder. Besides holding the abdominal viscera in place, these muscles bend the body forward, help to tilt the pelvis and are used in side bending and turning of the spine. They also help to force air out of the lungs.

The Knee Joint

The knee joint, halfway between the hip and the foot, is the largest joint in the body. It must support enormous weight and take the strain of most of the body's locomotion. You cannot sit, stand, walk, run or jump without the knees making a complicated motion. Yet the knee is very inefficiently made. It has a poor bone arrangement and is poorly held together. Once the ligaments that hold the knee are strained, there may be a weakness in the knee joint forever after.

I've noticed even in quite young children the tendency to bowlegs and knock-knees. Certainly, if your child has such a condition he should do corrective exercises and see an orthopedist.

Unfortunately, some children start out with good knee position and then, when they get to school, develop problems because they are given exercises that tend to cause hyperextension of the knee.

Don't allow your child to do exercises that require a locked, stiff knee. For example, many schools tell children to bend over and touch their toes, keeping their legs stiff. But the knee should always be bent slightly. In all standing exercises the pelvis should be tilted back, with the knees slightly bent. When your child (or anyone else) is bending forward to touch hands to the toes or the floor, the knees should bend, as the body bends.

The primary movement at the knee joint is to flex, or bend, and extend, or straighten. When in the standing position, the line of gravity of the body falls in front of the knee joint. Therefore, gravity tends to push the knee backward and to produce hyperextension and overstretch the hamstrings, the muscles at the back of the knee. The knees should not be pushed back and locked because the hamstring muscles will tend to lose their tension.

The position of the knee is dependent upon the position of the pelvis and leg bones. The knees are subject to great strain and stress in carrying the body weight around. If the weight of the child is not placed properly onto the legs, the knees will be thrown out of alignment and will develop improperly, becoming knocked or bowed.

This condition will throw the weight of the body in and onto the long arches of the feet, causing them to flatten.

The Foot

The feet bear the weight of the body. They push the body about, and with the use of the arches, make walking an easy, smooth movement by taking the jarring — like shock absorbers on a car. If the feet are not strong, the body will be thrown off balance and bad posture will result. Bad feet can cause bad posture. Bad posture can cause bad feet.

The elastic quantity of the muscles and ligaments of a strong foot holds up the arches and balances the foot The metatarsal, the arch at the base of the toes, flattens in standing and walking. It should spring back into an arch when it is not bearing weight. This arch must be kept strong so that the weight of the body will not break it down.

The other foot arch, the longitudinal, extends on the inside of the foot, from the heel to the base of the toes. It is not strong enough to bear the entire weight of the body, and children should not stand with their weight thrown on that arch.

To test the arches of a child's feet, take a piece of brown paper and place it on the floor. Wet the child's feet thoroughly and have him stand on the paper for a moment. If the arches are good the print on the paper will show the heel, the base of the toes, the ends of the toes, and the outside border of the foot from heel to toes. If more than that shows, you know your child has bad feet and should be given corrective exercises, both to strengthen the arches of the feet and to correct body posture.

The Daily Program

Every child, from the time he is five years old, ought to have a daily program of exercise to supplement his ordinary play activity. Even if you live in the country and your child seems to be climbing and jumping all day, he is using certain groups of muscles more than others and a well-planned exercise program is still important.

It isn't difficult to manage. You have, by this time, taught your child to brush his teeth and wash his face and perform other complicated health routines. Now teach him to exercise regularly. It is just as important if he is to avoid all the posture and pain problems that plague adults.

Don't scare the child by making the exercises seem too serious at the beginning. As we have seen, you can treat the exercises as games. Children love to imitate grownups and you'll find that the young child, especially, will be eager to join the fun of these routines.

Go to some part of your house which isn't cluttered with small pieces of furniture that will get in the way. If you have a fairly thick rug, you can exercise right on the rug. Otherwise, use a padded mattress cover or several thicknesses of blanket. Don't exercise on the hard bare floor. Don't use a plastic beach mat because it will stick to your body and feel very uncomfortable.

This book gives a daily program your child should follow after he has mastered all the exercises. It should take about twenty to thirty minutes a day. However, if your child seems to be getting tired or losing interest, break up the session into several small parts. With a five-year-old, start with one or two times per exercise, and as the child develops, work up to the number of times indicated at the bottom of each exercise. By age twelve, the child will be able to easily do each exercise the number of times recommended on the exercise page. Introduce only a few new exercises each day so that he isn't swamped with too much to learn. Don't move too quickly through the daily program, but give the child the chance to practice each new exercise until he has mastered it.

Don't overdo and make your child do more than the specified number of times per exercise or more than the daily program. No child should spend more than about a half-hour a day on exercises, and five- to eight-year-olds may be ready only for ten or fifteen minutes.

The exercises are arranged so that the child begins by lying down, moves to a sitting position and then stands up. This program is based on the structure of the spine and on the developing muscles.

(If the child has a special structural or muscle problem, work with the Exercises for Special Problems, pp. 73.)

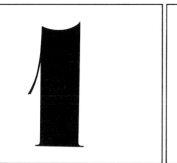

Rocking

When this exercise becomes easy to do, omit it and move to Exercise 2.

This exercise prevents sway-back.

1. Lie on the back on the floor, arms out at shoulder level, palms facing up. Bend the knees, with feet on the floor close to the hips. Put the feet together, with the soles on the floor. (There will be a space between the small of the back and the floor.)

2. Pull both knees up over the chest in a rocking motion. (The lower hips will lift slightly off the floor, and the small of the back will be pressed against the floor.)

3. Hold the small of the back against the floor and slowly lower one leg, placing the foot back on the floor close to the hips as in the starting position. Make sure the lower back is pressed against the floor.

4. Lower the other leg back onto the floor to the starting position, keeping the small of the back on the floor.

The Hip Rock

Do exercise 5 times.

This exercise prevents sway-back. It is the same movement as in the previous exercise, except that the feet remain on the floor.

1. Lie on the floor. Bend the knees and place the feet on the floor just below the hips, keeping feet together, soles on the floor, with knees apart. Place the arms out at shoulder level with the palms facing the ceiling.

2. Rock the hips back, pressing the small of the back against the floor. Hold for a count of three, with both feet on the floor, and the pelvis tilted backward. Relax.

The Tummy Tightener

Do exercise 5 times.

This exercise helps tighten the stomach muscles.

1. Same position as for Exercise 2, with knees bent, feet together on the floor, knees slightly apart, arms out at shoulder level with palms up.

2. Rock the pelvis backward, pressing the small of the back against the floor.

3. Hold the small of the back on the floor and pull the tummy in, down toward the floor. Hold for a count of four. Relax.

The Floor Arm Slide

Do exercise 5 times.

This exercise prevents "forward head and shoulders."

1. Lie on the back on floor, knees bent, feet on floor close to hips. Feet should be close together, knees slightly apart. Place fingers on shoulders. Pull elbows in close to the body. Push wrists and elbows down toward the floor. Rock pelvis backward with small of back on the floor. Pull tummy in. Push top of head away from the shoulders, but don't press chin against the neck. Keep head on the floor.

2. Turn palms so they face the ceiling. Keep wrists and elbows on the floor.

3. Slowly slide arms up over the head as far as possible. Only slide as far as the arms will go, keeping the elbows and wrists on the floor.

4. Slowly slide the arms back down to starting position. Keep elbows and wrists on the floor. Bring elbows in close to the body and bring fingertips to shoulders. Push waist back, pull tummy in. Push wrists and shoulders back toward the floor.

The Mule Kick

Do exercise 8 times
(4 for each leg).

This exercise strengthens the pelvic muscles and the muscles that rotate the thigh outward.

1. Lie on the floor, arms out at shoulder level with palms up. Knees bent with feet on the floor. Feet together with knees apart. Rock the pelvis backward, pressing the small of the back against the floor. Pull the tummy in tight.

2. With the stomach pulled in tight and the small of the back on the floor, bend one knee over the chest.

3. Straighten the leg up toward the ceiling. Turn the knee and foot outward.

4. Then slowly lower the leg down to the floor. The spine must stay on the floor with the tummy in tight all the time.

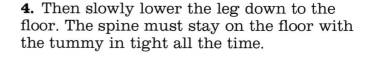

5. Bend the knee and place the foot back on the floor beside the other foot. The feet are together, side by side, pointing straight ahead. The knees are apart.

6. Bend the other knee over the chest.

7. Straighten the leg up toward the ceiling.

8. Turn the knee and foot outward and slowly lower the leg to the floor. Be sure that the small of the back does not come off the floor and that the tummy is pulled in tight. Bend the knee and place the foot back on the floor beside the other foot, feet close together with knees apart.

6

The Seesaw

Do exercise 5 times.

This exercise is for the abdominal muscles.

1. Lie on the floor, arms at sides, knees bent, feet together on the floor close to the hips.

2. Slide both feet down away from the hips as far as possible, trying to keep the soles of the feet on the floor, the knees slightly apart.

3. Raise the head off the floor.

4. Raise the shoulders and arms off the floor.

5. Raise the body up to the sitting position.

6. Reach down with fingers toward the feet. Keep knees bent and don't force the bend.

7. Slowly lower the body back to the floor — first your lower back touches the floor, then the upper back, then the shoulders, then the arms and finally the head.

Fingers and Toes

Do exercise 6 times
(3 for each leg).

This exercise is for the abdominal muscles.

1. Lie on the back on the floor, arms at sides, legs extended. Bend knees slightly and try to keep the bottom of the feet on the floor.

2. As you raise the head, bring the shoulders, back and arms up off the floor, all at the same time.

3. Raise one leg and touch both hands to the lifted foot.

4. Lower the back, arms, head and leg back to the floor at the same time. Keep the knee bent and place one foot on the floor beside the other foot. Rise again in the same manner and repeat the exercise with the other leg.

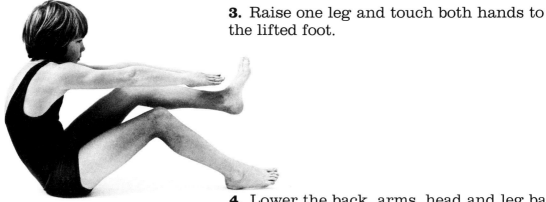

The Shrimp Kick

Do exercise 8 times on each side.

Exercises 8–12 strengthen growing muscles and help develop coordination and agility.

1. Lie on the floor on one side. Extend arms over the head as if you were about to dive. Rest the head on the underneath arm. Pull legs forward into a semicircular position, with knees bent slightly. (Tell the child he should curve like a shrimp.)

2. Raise the top arm and top leg up toward the ceiling. Let the knee bend and touch hand to foot.

3. Lower the arm and leg back down to the starting position.

9

The Knee Slap

Do exercise 10 times (5 times with each leg).

1. Lie on the back on the floor. Place the arms on the floor over the head, legs extended and together. Knees should be slightly bent.

2. Raise both arms and one leg, letting the knee bend. Touch hands to the raised knee.

3. Lower the leg and arms at the same time back to the floor, with the arms behind the head.

4. Raise both arms and the other leg, letting the knee bend. Touch hands to knee.

5. Lower the arms and leg back to the floor at the same time, arms behind the head.

10 Crossing Over

Do exercise 10 times
(5 times with each leg).

This exercise should be done in a continuous roll from side to side.

1. Lie on the back on the floor. Extend the arms out at shoulder level, palms up. Keep legs together and extended. Do not stiffen knees.

2. Raise one leg up and across the body to the floor on the other side. Let the knee bend. Touch toe to floor.

3. Bring the leg back to starting position.

4. Raise the other leg up and across the body to the opposite side. Touch toe to floor, allowing knee to bend.

5. Bring leg back to starting position.

11 The Bicycle

Do exercise 20 times.

1. Lie on the back on the floor, with arms extended at shoulder level, palms up. Place a rolled-up towel under the hips. Bend the knees over the chest. (You are going to ride a bicycle with your feet in the air.)

2. Straighten one leg up toward the ceiling.

3. Bend it back over the chest at the same time you straighten the other leg up toward the ceiling. Keep bicycling in a continuous, smooth movement, one leg after the other.

The Bicycle Backward

Do exercise 20 times.

Not every real bicycle can pedal backward, but that is what we want to do in this exercise.

1. Get in the same position as for the previous exercise.

2. Begin with the leg going down.

3. Then bring leg up, and back to chest in reverse motion.

13 Wall Sitting

Do exercise 3 times.

This is the basic position for the wall exercises for preventing "forward head and shoulders."

1. Sit on the floor close to the wall. Cross the legs tailor-fashion, with hands on the floor beside the hips, with palms up.

2. Bend forward from the hips as far as possible.

3. Wriggle hips backward until they are completely against the wall. Then slowly raise the body back up, one vertebra at a time, until the whole spine and head are against the wall.

4. Push the small of the back against the wall. Pull tummy in. Push shoulders back toward wall. Pull shoulders down. Push top of head toward ceiling, keeping chin at right angle. Hold for count of ten.

When the child can sit tailor-fashion with the whole spine against the wall, this exercise can be omitted.

Head Turns

Do exercise 8 times
(4 times to each side).

This exercise helps strengthen the neck muscles that hold the head in the proper position.

1. Sit on the floor with the hips, back and head against the wall. Cross the legs tailor-fashion. Place hands on the floor beside the body with palms up. Push the small of the back against the wall by rocking hips backward.

Pull shoulders back toward the wall and down toward the floor.

Keep chin at right angle to neck and push top of head toward ceiling.

2. Now, keeping the head against the wall, slowly turn the head to one side as far as possible, with the shoulders back and down, the top of the head pushing up.

3. Turn back to center.

4. Then turn head to other side.

5. Turn back to center.

15 Head Bends

Do exercise 8 times
(4 on each side).

This exercise strengthens the neck muscles.

1. Take the same position as for head turns in the previous exercise. The small of the back is on the wall, shoulders are back against the wall, pulling down toward the floor. The chin is at a right angle to the neck; hands are on the floor, palms up.

2. Keeping the head against the wall, slowly bend the head to one side, bringing the ear down toward the shoulder.

3. Bring the head back up. Push the top of the head toward the ceiling. Shoulders are kept back and down. Waist is against the wall.

4. Slowly bend the head to the other side, bringing the ear toward the shoulder.

5. Bring the head back up. Push top of head toward ceiling. Keep shoulders back and down.

16

Climbing the Wall

Do exercise 4 times.

This exercise helps prevent sway-back and teaches the correct pelvic tilt for the standing position.

1. Stand with the back and head against the wall, feet about 4 inches away from the wall and about 4 inches apart, arms at sides. Keep the feet parallel to each other. Do not turn toes out.

2. Bend knees to an angle of about 40 degrees, as in illustration.

3. Then bend forward from the waistline, keeping hips against wall. Let arms drop forward as far as they will comfortably go.

4. Slowly raise the body, pressing the spine against the wall, one vertebra at a time. Hold this position, pull the shoulder blades back toward the wall and pull the shoulders down toward the floor. Relax.

When the child can do this exercise easily, omit it from the program and do Exercise 17.

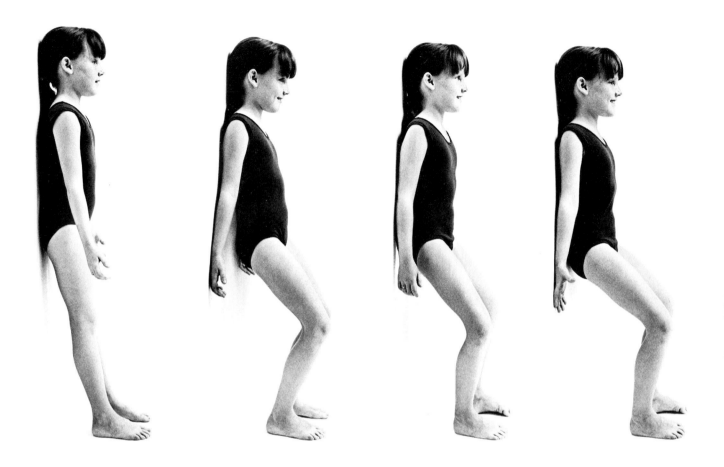

This exercise strengthens the muscles that hold the legs and the pelvis in the correct postural position.

1. Stand with the head and back against the wall, with the feet 4 inches away from the wall and 4 inches apart, parallel to each other. Arms are at sides, palms facing out.

2. Bend knees to an angle of about 40 degrees, as in illustration.

3. Pull knees apart, but keep toes pointing straight ahead.

4. Rock the pelvis back so that the small of the back is pressing against the wall and the stomach is pulled in. Push shoulders toward wall. Pull shoulders down toward floor. Hold to a count of five. Relax.

Lining Up the Spine

Do exercise 4 times.

This exercise accentuates the correct standing posture and strengthens the muscles.

1. Stand against the wall, feet 4 inches from the wall and 4 inches apart, arms at sides. Bend knees slightly. Pull knees apart. (You will notice that the inside arches of the foot lift up off the floor.) Push the small of the back against the wall. Pull tummy in. Push the shoulders back toward the wall, pulling them down. (Shoulders will come slightly forward as they pull down.) Push the top of the head toward the ceiling, the chin at a right angle to the neck.

Standing Straight

Do exercise 4 times.

2. Now slide the body up the wall as far as possible without letting the small of the back come off the wall. Hold.

Correct Standing
Position

Step away from the wall, feet a comfortable distance apart and pointing straight ahead.

Bend knees slightly — the tiniest bit.

Rock the pelvis back, pulling hips under, but slightly, as if you were going to sit.

Keep arms loose and relaxed at sides. Pull shoulders back and down. (Arms will come slightly forward.)

Lift the chest.

Push the top of the head toward the ceiling. (The weight of the body is on the heels, the outside borders and the balls of the feet.)

The next three exercises require a wand* about 30 inches long.

Side Bending

Do exercise 16 times (8 times to each side).

This exercise is for the antigravity muscles which hold the body upright.

1. Stand in the center of the room. Separate legs comfortably, toes pointed straight ahead. Bend knees slightly and turn them outward. Pull hips under, in a slight sitting position. Push the top of the head toward the ceiling. Take the wand and hold it at each end, down in front of the body, arms straight.

2. Raise wand over head, keeping arms straight and parallel with ears.

3. Bend from the waist to the right as far as possible, keeping elbows beside the ears.

4. Come back up, with elbows beside ears; keep hips pulled under.

5. Bend to the other side.

6. Come back up and bring the wand to the starting position.

*Your hardware store can cut doweling to this size. The wand should be very lightweight. If your child is very young and likely to poke himself, look around for a plastic baton or any other wand with protected ends.

The Waist Twist

Do exercise 8 times (4 times to each side).

This exercise will strengthen the muscles that hold the body upright.

1. Stand in the center of the room as in Exercise 19, holding the wand down in front of the body, legs apart, knees bent. Pull knees apart. Pull hips under (sit slightly). Hold head up; push the top of the head toward ceiling.

2. Raise wand over head, with the hips facing forward as much as possible.

3. Twist the upper body, head and chest to one side.

4. Twist the body to the front.

5. Twist to the other side.

6. Twist to the front. Make sure hips stay facing front and arms stay overhead.

This exercise is for the muscles that hold the body upright.

1. Stand as in Exercises 19 and 20, bend knees, pull knees outward. Pull hips under. Raise wand over head, with elbows beside ears.

Down and Up

Do exercise 8 times.

2. Bring the wand over and down to the floor, bending the knees as you do so.

3. Bring the wand back over the head with the knees slightly bent and the hips pulled under. Keep elbows beside ears. Don't let the head come forward in this position.

This exercise is for coordination and balance.

1. Stand in the center of the room, with legs about 4 inches apart. Pull hips to a slight sitting position. Pull tummy in. Keep the back straight. Bend the knees slightly and turn them outward.

Hold a wand with one hand at each end.

2. As you raise the arms up over the head, jump up at the same time. Land with legs far apart. When you land, knees should be bent and hips pulled under. (Never let your child land with stiff knees.)

3. Jump up again and land with legs closer together, only 4 inches apart. Bring arms down at the same time. Continue jumping, landing first with legs far apart, then closer; arms up, then down.

A Better Jumping Jack

Do exercise 10 times until it's easy, then increase to 20.

The Tightrope Walk

Children may do this exercise as much as they like.

This exercise will improve balance.

Find a straight line on the floor, either the edge of a rug or a line on the floor in the pattern of your linoleum. The line should run at least 10 to 12 feet.

1. Stand on the line with arms extended at shoulder level. Place one foot in front of the other on the line. Begin to walk like a tightrope walker, making sure the heel comes down first and then the ball of the foot.

2. Walk about halfway down the line, then slide the forward leg forward right on the line. Bend your back knee down and place it on the line.

3. Stand up, bring the back leg forward and continue balancing down the rest of the line.

This exercise strengthens the muscles of the upper back.

1. Stand in the center of the room with legs apart. Bend knees slightly and turn them outward. Pull the hips under in a slight sitting position. Pull tummy in, with arms at sides. Keep the back straight and the head up tall.

2. Now raise one arm forward, up, back and down in a large circle. Do the same with the other arm. Keep arms going, one arm and then the other, in a continuous smooth movement.

24

The Backstroke

Do exercise 20 times (10 times with each arm).

25

The Windmill

Do exercise 8 times (4 in each direction).

This good general exercise will help almost all the muscles of the body. You will be making a large circle with your arms in a continuous smooth movement.

1. Stand in the center of the room, legs apart, knees bent, hips tucked under.

2. Raise your arms over your head and clasp your hands.

3. Bending your knees as you bend, swing your arms first to your right toe.

4. Then swing your arms to your left toe.

5. Then swing your arms way back over your head to the starting position.

This exercise strengthens the leg and back muscles and is good for coordination and balance.

1. Stand in the center of the room, feet slightly apart. Remember that the knees are bent, hips pulled under.

2. Bend your knees as you bend down and place your hands on the floor in front of your feet.

The Cheerleader

Do exercise 4 times.

3. Spring up and jump in the air, raising your arms overhead like a cheerleader.

4. Be sure that your knees are bent when you land.

The Donkey

Do exercise 4 times.

This exercise strengthens the arm muscles.

1. Place your hands and feet on the floor, knees bent under the body.

2. Extend your right leg back, keeping your left leg bent under you.

3. Spring up and change the position of the legs.

4. Bring the right leg forward and the left back.

28 The Grasshopper

Do exercise 10 times.

This exercise strengthens the leg muscles.

1. Take the same position as for the previous exercise.

2. Straighten your right leg back as far as you can without letting the small of the back drop.

3. Straighten the other leg back to the same position.

4. Bring the right leg to the original position.

5. Bring the left leg to the original position.

Exercises for Special Problems

If your child has special structural or muscle problems, concentrate on those parts of his body which need help. If you notice, for example, that he has knock-knees or bowlegs, stress the exercises for that condition in the daily program.

The exercises in the following section are necessary only for children who have exaggerated postural faults. There is no need for children without special problems to do them. A child with special problems should do all the regular exercises in the daily program plus those that follow which pertain to his particular problem. Bear in mind that all children with such problems should see an orthopedist.

The Two-Leg Slide

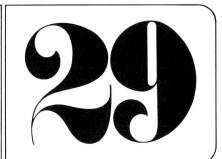

Do exercise 4 times.

This exercise helps prevent and correct sway-back, bowlegs and knock-knees, and also strengthens the arches.

1. Lie on the back with the arms out at shoulder level, palms up. Bend the knees and place the feet on the floor, keeping feet together, soles on the floor, with knees apart. Rock the pelvis backward so that the small of the back is on the floor. Pull tummy in tight.

2. Keeping the spine on the floor, slowly slide both feet down away from the hips. Keep feet side by side, with big toes and heels touching each other. The knees should pull outward. Slide the feet down only as far as you can without the spine coming off the floor. Don't let the back arch. (Big toes, heels and outside borders of the feet are on the floor. The inside arches pull upward and outward.)

3. Slowly slide one leg back to starting position. Make sure the knee is turned outward.

4. Slowly slide the other leg back.

30

The One-Leg Slide

Do exercise 8 times
(4 times for each leg).

This exercise helps correct sway-back, knock-knees and bowlegs, and strengthens the arches of the feet.

1. Start in the same position as for the previous exercise: on the back with arms out at shoulder level, palms up. Knees bent, feet on the floor close to the hips, knees slightly apart. Rock the pelvis back so that the small of the back is on the floor. Pull the tummy in tight.

2. Slowly slide one foot down away from the hips. Keep toes pointing straight ahead. Don't let the feet turn out. Keep knee turned outward.
  Slide the foot down only as far as you can without the spine coming off the floor. Toes, heel and outside border of the foot should stay on the floor. (As the foot slides down, you will notice the arch on the inside of the foot is being pulled upward and outward.)

3. Slowly slide the foot back to the starting position, being sure to keep the knee turned outward, the spine on the floor and the tummy in.

4. Slowly slide the other foot down in the same manner.

5. Slowly slide the foot back to starting position.

31

The Bowlegged Slide

Do exercise 8 times (4 for each leg).

This exercise corrects knock-knees and bowlegs.

1. Sit on the floor, hands behind the body, palms down on the floor for support. Bend knees, placing soles of the feet together, close to the body. Rock the pelvis back and pull tummy in.

2. Slowly slide the feet down away from the body as far as possible, keeping the soles of the feet together, knees pulling outward.

3. Now, slowly bring one leg back. Keep the knee pulling outward.

4. Slowly bring the other leg back until the soles of the feet touch again.

Funny Faces

Do exercise 6 times.

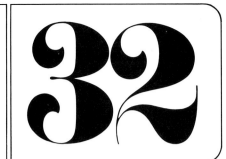

This exercise corrects knock-knees and bowlegs.

1. Sit on the floor, with hands behind the back for support. Extend legs in front and bend the knees slightly. Keep feet close together. Take a crayon and draw a face on each knee.

2. Turn the faces outward, away from each other as far as possible. <u>Keep knees together</u> while you do this. Do not let them separate at any time. (As the knees turn outward, you'll notice the calves will turn in.)

3. Now turn the faces toward each other.

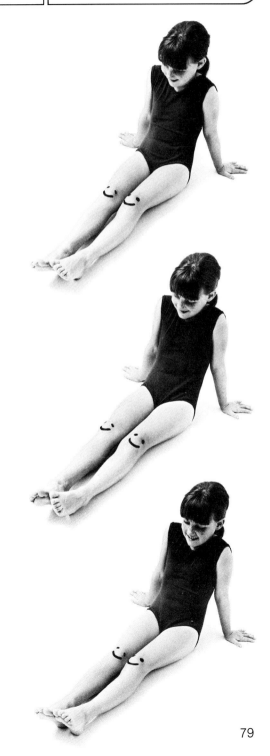

This exercise corrects knock-knees and bowlegs.

1. Stand with the back and head against the wall, feet about 4 inches away from the wall, 4 inches apart and parallel. Do not turn toes out. Keep arms at sides.

33

Sliding Up the Wall

Do exercise 6 times.

2. Bend knees deeply.

3. Push the small of the back against the wall. Pull shoulders back toward the wall. Pull downward with the fingers. Push the top of the head up toward ceiling.

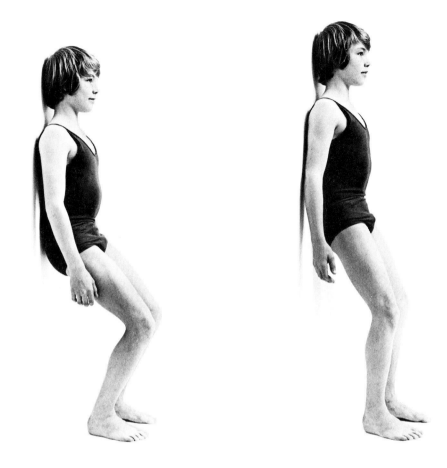

4. Now, pull the knees outward and apart as far as possible. Do not let toes turn out. Keep them pointing straight ahead. (The inside long arches will pull up and outward.)

5. <u>Slowly</u> slide the body up the wall, pulling the knees outward as they straighten. Slide up the wall only as far as possible with the small of the back staying on the wall and knees bent, pulling outward.

34

The High-Stepper

Do exercise 12 times (6 times with each leg).

This exercise corrects knock-knees and bowlegs.

1. Stand with the back and head against the wall, feet close together, 2 inches from the wall, toes pointing straight ahead. Bend the knees slightly and turn them outward. Push the small of the back against the wall. Pull tummy in. Pull shoulders back and down. Pull downward with the fingers. Push the top of the head up toward the ceiling.

2. Hold this position and shift the weight to the left foot. Now bend the right knee up toward the chest, pulling the knee outward.

3. Return the foot to the floor.

4. Shift the weight to the right foot and bend the left knee up toward the chest, pulling the knee outward.

5. Return the foot to the floor.

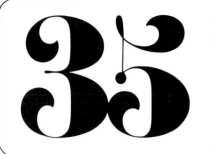

Tailor on the Wall

Do exercise 8 times (4 times with each arm).

This exercise corrects "forward head and shoulders." Do it only if your child has a problem of this type.

1. Sit on the floor with buttocks and back against the wall, legs crossed tailor-fashion. Place fingertips on shoulders. Bring elbows in close to the body and wrists and elbows back against the wall. Push the top of the head toward the ceiling. (Chin is at right angle to neck.)

2. Turn the palm of one hand facing outward.

3. Push hand against the wall and slowly slide the arm along the wall up over the head, as far as possible, with wrists and elbows on the wall. Keep the other arm back against the wall.

4. Slowly slide the arm back down and place fingertips on shoulder. Keep elbows and wrist on the wall all the time. Bring elbow in close to body. Pull shoulders down. Push top of head toward ceiling.

5. Turn the other palm out.

6. Slowly slide the arm along the wall up over the head as far as possible with the wrist and elbow on the wall. Keep the other arm back on the wall as you slide this arm up.

7. Slowly slide the arm back down and place the fingers on the shoulder. Push wrists back toward wall. Push shoulders down toward floor. Push top of head toward ceiling.

36 Two Tailors on the Wall

Do exercise 5 times.

This is another exercise for "forward head and shoulders." Use it only if your child has a problem of this type.

1. Sit on the floor with hips, head and back against the wall, legs crossed tailor-fashion. Place fingertips on shoulders, with wrists and elbows against the wall, elbows in close to the body. Be sure the small of the back is against the wall all of the time. Push the top of the head toward the ceiling.

2. Turn palms outward as in the previous exercise, pushing hands against the wall.

3. Slowly slide both arms along the wall up over the head as far as possible with the elbows and wrists against the wall.

4. Slowly bring the arms back down the wall. Keep arms on the wall all the time. Place fingertips on shoulders. Press elbows and wrists back against the wall, press the small of the back against the wall. Push the top of the head toward the ceiling.

Up Against the Wall

Do exercise 6 times (3 times with each arm).

This exercise is for "forward head and shoulders." Do it only if your child has a special problem of this type.

1. Stand with back and head against the wall, with feet 4 inches away from the wall and 4 inches apart. Keep toes parallel, pointing straight ahead. Bend knees slightly, to the extent indicated in illustration. Pull knees apart, keeping feet straight ahead. Rock the pelvis back and push the small of the back against the wall. Place fingertips on shoulders. Bring elbows in close to the body. Push wrists and elbows back against the wall.

2. Turn one palm outward and slowly slide the arm up the wall over the head as far as possible with elbow and wrist on the wall. Hold it there.

3. Now turn the other palm out and slowly slide the arm on the wall up over the head as far as possible, keeping the elbow and wrist on the wall.

4. Slowly bring one arm back down, keeping it against the wall. Place fingertips on the shoulder, bring elbow in close to the body, push elbow and wrist against the wall.

5. Slowly bring the other arm back down. Place fingertips on shoulders, push wrists back, elbows back, against the wall. Push the top of the head up toward the ceiling.

38

Holding up the wall

Do exercise 5 times.

This exercise corrects "forward head and shoulders."

1. Stand against the wall. Bend the knees slightly and pull them apart. Keep feet parallel, pointing straight ahead. Arms at sides. Rock the pelvis back, pressing the small of the back against the wall. Place fingertips on shoulders, elbows close to the body. Push top of head up toward ceiling.

2. Turn palms out and slowly slide both arms up the wall, keeping wrists and elbows against the wall. Keep small of back on the wall.

3. Slowly bring both arms down, keeping wrists and elbows against the wall. Place fingertips on shoulders. Bring elbows in close to wall.

Check: Waist back on the wall, top of head up toward ceiling, shoulders back and down.

This exercise strengthens the muscles that hold you upright.

1. Stand against the wall, feet 4 inches away from the wall and 4 inches apart. Keep the feet parallel, pointing straight ahead. Bend the knees and pull them apart. Push the small of the back against the wall. Raise arms up above the head. Clasp hands and press them against the wall.

2. Bend the body over to one side as far as possible, keeping waist, back and arms against the wall.

3. Come back up to starting position, keeping the small of the back glued to the wall at all times. Repeat exercise, bending to the other side.

39

Wall Bends

Do exercise 4 times (2 to each side).

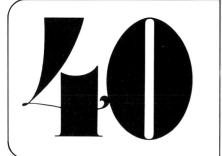

The Fan-Fan

Do exercise 8 times.

This exercise corrects pigeon-toes.

1. Sit on the floor, legs extended in front, knees and feet together, arms extended back, with hands placed on the floor, palms down for support. Keep knees and heels touching.

2. Pull toes up toward the knee.

3. Turn feet outward, being sure that as the feet turn outward the knees do not separate. (This is a little hard but is very good for the pigeon-toed child.) Do not stiffen the knees. They should be slightly bent.

4. Bring feet together.

This exercise is similar to the one for flatfeet on page 17, but here, for pigeon-toed children, the toes are pulled outward instead of inward.

You must have a straight line running at least 10 or 12 feet on the floor. You can use the edge of a rug, or a line on the rug or on linoleum.

1. Walk placing the feet on either side of the line, parallel with it. Walk with heels touching the floor first.

2. Before the rest of the foot touches the floor, turn pigeon-toed foot outward so that it is straight and parallel with the line. Then place the outside border of the foot on the floor beside the line.

3. Place the ball of the foot and then the toes on the floor. Be sure the knee bends in a straight line over the foot, not inward.

Walking a Line

(For pigeon-toes)

Do exercise a few times a day.

If your child has flatfeet, use the following exercise as well as those which appeared earlier in the section on the toddler: Building a Mountain (p. 15), Picking Up Marbles (p. 18) and Walking a Line for Flatfeet (p. 17).

The Clown Walk

Do until feet get tired.

This is another exercise that strengthens the two arches of the foot. It is particularly good for children with flatfeet, but it won't hurt anyone to do it.

Walk barefoot around a carpeted room with knees bent and turned outward. Curl toes in and walk on the outside edges of the feet.